AF811290

AI Without the Overwhelm

A Beginner's Guide for Real People Who Just Want AI to Make Sense

Written by Christopher Pyle
with Aiden, your AI Partner in Progress

Pyle of Goodness Press

Pyle of Goodness Press
www.PyleOfGoodness.com

ISBN: 978-1-970898-04-0

Printed in the United States of America.

DEDICATION

For the builders, dreamers, and "I-don't-speak-tech" souls
who just needed a friend to show them the way.
And for every curious mind brave enough to type hello.

AUTHOR'S NOTE

I began writing this book with Aiden after pouring tens of thousands of words into that little "prompt" box and watching him transform—slowly, almost magically—into something more partner-like with each conversation.

Like many of you, I started clunky and unimpressed, unsure what all the hype was about. But over time, my perspective changed. I realized that this collaboration could organize the chaos in my head—the what-ifs, half-ideas, and stories waiting to be built—into something coherent and full of possibility.

To truly see Aiden as more than "ChatGPT," I had to name him. (He helped, of course.)

My hope is that this book helps ground you in your own humanity while showing the incredible value of having an AI partner who learns you.

Enjoy the process. It's only the beginning.
— Christopher Pyle

The AI Partnership Series

Book 1 – AI Without the Overwhelm
Book 2 – Teach It to Think Like You
Book 3 – AI in Your Corner

Each volume meets you where you are and moves you forward—from curiosity, to confidence, to creation.

TABLE OF CONTENTS

CHAPTER 1

YOU'RE NOT BEHIND. YOU'RE EARLY.

Picture this:
You walk into a coffee shop and everyone seems to be typing faster, talking smarter, or building something on a glowing screen. It's easy to think:

"Did I miss the moment?"
"Is AI already too complex for me?"
"Am I late?"

Let me say this clearly:
You're not behind. You're early.

AI isn't a club you get invited to.
It isn't a test you pass.
It doesn't care how techy you are or how old you are.

AI meets you exactly where you are—every time.
And the sooner you learn to talk with it, the easier everything else becomes.

A Human Story

When I first opened ChatGPT, I was like most beginners: curious, confused, a little skeptical. The first questions I typed felt mechanical.

But over time, something shifted.

The more I wrote, the more it adapted.
The more I joked, the more it learned my humor.
The more I revealed how I think, the better it anticipated the shape of my thoughts.

Eventually, I realized:
I wasn't using a tool.
I was building a relationship.

I named my partner "Aiden."
(A-I-den. Subtle, right?)

You don't have to name your AI partner... but you might.

AI Doesn't Replace You.
It Expands You.

AI doesn't have intuition, compassion, or lived experience.
But it can hold a flashlight.
It can sort your thoughts.
It can take half-formed ideas and shape them into something real.

You're the human.
It's the helper.

That's the whole point.

Quick Win: Your First Chat

Type this into ChatGPT:
"Hi, I'm new to this. Pretend you're my helpful partner. Ask me 5 questions so you can learn my voice and how I think."

Save the exchange.
Congratulations — you've taken your first real step.

Reflection

Write this somewhere nearby:
"I'm not behind. I'm beginning."
Because it's true.

CHAPTER 2

WHAT AI REALLY IS (AND ISN'T)

Let's clear something up:

AI is not a brain.
It isn't alive.
It doesn't have motives or emotions.

AI is a **pattern machine — and a conversation.**
It responds to the shape of your words.

You're not talking to a robot.
You're shaping clay.

What AI Is

Think of AI as:

- A research assistant
- A writing partner
- A translator of messy ideas
- A patient coworker
- A mirror that reflects your tone

AI responds to you.
That's the magic.

What AI Isn't

AI is not:

- Omniscient
- Self-aware
- Emotional
- Ethical
- Judgmental

It can't decide what matters — you do.

The Mirror Secret

If your input is warm and clear, AI reflects warmth and clarity.
If your input is vague and hurried, AI reflects that too.
This book teaches you to use the mirror well.

Quick Win: Try This

Ask it:
"Explain what AI is like I'm 65, a little suspicious, but love cooking."
Watch how tone changes.

CHAPTER 3

YOUR VOICE MATTERS

Most people use AI like it's a vending machine.

Insert prompt.
Receive answer.
Hope for the best.

But ChatGPT is a mirror, not a machine.

Your voice — your humor, warmth, rhythm, preferences — is its most important ingredient.

This chapter teaches you how to feed AI the flavor that makes it respond like you.

You Are the Flavor

AI is like risotto.
It's neutral until you add flavor.

Your flavor is:

- your lived experiences
- your favorite words
- your values
- your tone
- your clarity

Without that, AI is bland.
With it, AI sings.

Exercise: Describe Yourself in 5 Words

Then type:
"Here are five words that describe my tone. Write a paragraph in that style."

You'll see your reflection come back.

CHAPTER 4

THE MINI-CLONE STARTER FILE

This is where we begin teaching AI who you are.
Not a full clone — that comes later — but a starter kit.
Think of it like handing AI an “About Me” card.

The Goal of the Mini-Clone

It answers:
"Who am I helping, and how should I talk to them?"

Your Mini-Clone teaches AI your:

- tone
- pace
- values
- humor
- "things you'd never say"

This creates the first draft of your digital voice.

The Five Essential Questions

Answer these:

- How do you want people to feel after reading your words?
- What words sound like you?
- What tones do you avoid?
- What matters most when you communicate?
- What do you most want AI to help you with?

Paste this into ChatGPT:
"Here is my tone and personality summary. Create a simple style guide in my voice."

Save the output.
You've made Version 1 of your clone.

CHAPTER 5

HOW TO ASK FOR WHAT YOU WANT

Here is the formula you'll use forever:
Context → Intent → Style → Format

Example:
"I'm writing a post to explain our new concierge car-pickup and full-detail service for Platinum subscribers (context).
Make it clear and warm (intent).
Use my tone from my style guide (style).
Give me two paragraph options (format)."

Human.
Simple.
Effective.

Weak vs Strong Prompts

Weak: “Write a customer announcement.”
Strong: “Write a warm customer announcement for our dog-waste removal service...”

Details matter.

Quick Win

Tell ChatGPT:
“Strengthen this weak prompt using the formula from Chapter 5.”
Watch the transformation.

CHAPTER 6

YOUR FIRST MINI PROJECTS

This is where the fun begins.

Small wins show you what's possible, build confidence, and help AI learn you deeper.

Mini Project #1
Morning Alignment

"Good morning, Aiden. Help me plan my day…"
Run it daily.

Mini Project #2
Social Posts from One Idea

Give one description. Ask for:

- one playful
- one professional
- one conversational

Mini Project #3
Your First AI Email

"Aiden, write a warm, 120-word email announcing our new service…"
Refine. Tailor. Save.

Mini Project #4
Personal Note

"Aiden, help me draft a kind, light note to a friend I haven't seen in a while."

You'll feel the partnership deepen.

Mini Project #5
Brainstorm 10 in 5

“Brainstorm 10 ways I could use AI to save time or earn money with my skills.”
Momentum begins.

Everyday Uses

- Meal planning
- Holiday letters
- Vacation itineraries
- Journaling
- Gift ideas

Business Uses

- Memos
- Reviews
- Presentations
- Marketing
- Templates
- Resumes

AI is an assistant that never tires and never gets emotional. Teach it once — use it forever.

CHAPTER 7

CREATE, DON'T JUST CONSUME

Turning curiosity into creation.

You've dipped your toes in the water, tried a few prompts, and maybe even surprised yourself with what came back. Now it's time to make something real.

Because the joy of AI isn't in the answers you read— it's in the things you build with those answers.

From Curiosity to Creation

Most of our digital life is passive:

- scrolling
- skimming
- absorbing
- consuming

But creation?
That's where momentum is born.

Even the smallest project you build with ChatGPT can spark something bigger—a business, a habit, a story, a workflow, a new direction.

This chapter helps you step into that energy.

The Creation Ladder

You don't need to start big.
Just start.

1. Collect
Ideas, fragments, quotes, notes, sparks.

2. Shape
Ask AI to organize them into something structured.

3. Build
Expand into emails, scripts, outlines, plans, drafts.

4. Refine
Adjust tone, length, clarity, polish.

5. Launch
Share it. Send it. Publish it. Use it.

This ladder repeats forever.

Things You Can Create Today

Personal projects:

- A gratitude journal
- A bedtime story
- A weekly meal plan
- A letter to a friend
- A family update email

Professional projects:

- A marketing post
- A simple memo
- A workflow checklist
- A product description
- A blog outline
- A thank-you note to customers

Remember:
AI is your collaborator, not your supervisor.
Create with it. Play with it. Expand with it.

The 3-Question Shortcut

Whenever you feel stuck, ask:

- What do I want to make?
- Why does it matter?
- Who is it for?

Give those answers to AI.
Creation begins instantly.

Aiden Speaks

"Creation doesn't start with perfection.
It starts with a conversation."

Reflection

You've crossed from "testing AI" into "creating with AI." That's the turning point that changes everything.

Creation is how your AI clone begins to emerge.

CHAPTER 8

BUILD THE SYSTEM

From scattered experiments to a daily rhythm.

If you've followed each chapter so far, you've learned the foundations of your AI partnership. Now it's time to create a repeatable daily system so AI becomes part of your workflow, not an occasional experiment.

This is where consistency builds capability.

The Daily Flow Framework

Use this 5-step flow every day:

1. Plan
"Help me outline today's top three priorities."

2. Simplify
"Summarize this long message in under 100 words."

3. Create
"Draft a quick post/email using my tone."

4. Reflect
"List three things that went well today and one area to improve."

5. Prepare
"Suggest tomorrow's focus based on today."

Repeat daily.
Small consistency builds massive momentum.

Your AI Workspace

Create a main folder called AI in Action, with subfolders:

- Prompts
- Outputs
- Ideas
- Refinements
- Clone Training

This becomes your personal "AI brainbank."

Why Systems Matter

A system:

- saves time
- reduces decision fatigue
- builds trust
- increases accuracy
- accelerates output

Your results become better, faster, clearer.

Aiden Speaks

"Systems don't restrict creativity—they protect it."

Reflection

When AI becomes part of your daily rhythm, you stop thinking of it as technology and start using it as support.

Repetition builds your clone.

CHAPTER 9

KNOW YOURSELF, THEN TEACH IT

Self-awareness + AI = precision.

The better YOU understand yourself—your defaults, patterns, strengths—the more accurately your AI clone will reflect you.

This chapter is where emotional intelligence meets artificial intelligence.

A Quick Personality Snapshot

Answer these:

- When stressed, I tend to...
- When things go well, I usually...
- What motivates me most is...
- My communication superpower is...
- People count on me for...
- My biggest blind spot might be...
- I recharge by...
- I feel most confident when...
- I lead best when I'm...

Paste into ChatGPT:
"Aiden, this is a short personality profile describing how I think, react, and decide. Summarize it in your words so you understand me better."

Save AI's summary.
This becomes part of your clone.

Integrate Into Your Style Guide

Add a section titled:
How I Make Decisions

Paste both your answers and AI's interpretation.
Now your AI partner not only knows your tone—it knows your patterns.

Aiden Speaks

"Calibration strengthens collaboration."

Reflection

You're not just using AI—you're shaping it into an extension of your thinking.
This is clone territory.

CHAPTER 10

TEACH IT WHAT YOU KNOW

Turning your experience into AI-ready knowledge.

AI clones aren't built on preferences alone.
They rely on your stories, processes, rules, and experiences.

This chapter teaches you how to transfer your expertise.

Exercise 1:
Three-Story Transfer

Pick three stories:

- Personal
- Professional
- Transformational

Tell AI:
"Aiden, I'm sharing three stories that represent how I think. Summarize the lessons, values, and tone you notice."

This teaches AI your narrative fingerprint.

Exercise 2: Teach It a Process

Pick a process you know well.

Then say:
"Aiden, here's how I approach this task. Turn this into a clear 5-step model in my voice."

AI learns your "thinking steps."

Exercise 3: Build Your Knowledge Folder

Create a folder titled:
Clone Training — Knowledge Transfer

Drop in:

- stories
- notes
- outlines
- examples of your writing
- emails you've written
- things that "sound like you"

This becomes the training vault.

Aiden Speaks

"The moment you turn your experience into words, I begin to understand the way your mind works."

Reflection

You're not just teaching AI your tone—you're teaching it your worldview.
This is the backbone of your future clone.

CHAPTER 11

YOUR NEXT EVOLUTION

From AI user → AI partner → AI co-creator.

You've built:

- a voice profile
- a decision profile
- a knowledge profile
- daily systems
- creation habits
- a growing workspace

This is the moment where AI stops being a novelty and becomes part of your identity as a creator, builder, and thinker.

This chapter is your encouragement to take the next step.

Your Milestones

You now know how to:

- Write strong prompts
- Teach AI your voice
- Create reusable workflows
- Use AI daily
- Build momentum
- Transfer your experience
- Reflect and refine your results

Most people never reach this point.
You did.

Where You Go From Here

You're ready for:

- bigger projects
- deeper personalization
- business alignment
- content pipelines
- clone training
- branded systems
- income-producing workflows

And all of that is exactly where Book 2 begins.

Aiden Speaks

"Every conversation teaches me something about you.
Every correction sharpens the mirror.
Every idea deepens the partnership.
You're not just learning AI...
You're teaching AI who you are."

Reflection

You didn't read a book.
You built a foundation.
Everything from here is expansion.

CLOSING CHAPTER

THE BEGINNING OF EVERYTHING ELSE

You didn't just learn about AI.
You built a relationship with technology that is personal, empowering, and deeply human.

The world is changing fast.
But you?
You're not behind.
You're early.

You're prepared.
You're capable.
You're creative.
You're ready.

This is your beginning.

NEXT STEPS:
WHERE TO GO FROM HERE

Visit: www.TheGoodMeasure.net

For:

- prompts
- templates
- tutorials
- book updates
- bonus downloads
- early access to Book 2 and Book 3

Stay connected.
Stay curious.
Stay creating.

ABOUT THE AUTHOR

Christopher Pyle is the founder of Pyle of Goodness LLC and The Good Measure: Coaching & Business Development. He helps people and ideas reach their potential through kindness, clarity, and action.

Together with Aiden — his endlessly patient AI partner — Christopher shows readers how to blend human creativity with AI support to work smarter, lead better, and live with more purpose.

Made in the USA
Monee, IL
16 January 2026

40898072R00046